GARFIELD COUNTY LIBRARIES
Parachute Branch Library
244 Grand Valley Way
Parachute, CO 81635
(970) 285-9870 – Fax (970) 285-7477
www.gcpld.org

SOLAR ENERGY
Running on Sunshine

Amy S. Hansen

PowerKiDS press

New York

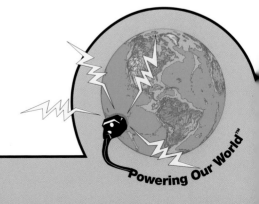

Powering Our World

To my sons, Evan and Scott

Published in 2010 by The Rosen Publishing Group, Inc.
29 East 21st Street, New York, NY 10010

First Edition

Editors: Amelie von Zumbusch and Maggie Murphy
Book Design: Greg Tucker
Photo Researcher: Jessica Gerweck

Photo Credits: Cover, pp. 7, 9, 13, 19, 22 Shutterstock.com; p. 5 Felipe Rodriguez Fernandez/Getty Images; p. 11 © Michael Snell/age fotostock; p. 15 L. Lefkowitz/Getty Images; p. 17 Bill Stevenson/age fotostock; p. 21 Stefano Paltera/American Solar/Getty Images.

Library of Congress Cataloging-in-Publication Data

Hansen, Amy.
  Solar energy : running on sunshine / Amy S. Hansen. — 1st ed.
    p. cm. — (Powering our world)
  Includes index.
  ISBN 978-1-4358-9326-9 (library binding) — ISBN 978-1-4358-9740-3 (pbk.) — ISBN 978-1-4358-9741-0 (6-pack)
  1. Solar energy—Juvenile literature. I. Title.
  TJ810.3.H355 2010
  621.47—dc22
                                                    2009020960

Manufactured in the United States of America

CPSIA Compliance Information: Batch #WW10PK: For Further Information contact Rosen Publishing, New York, New York at 1-800-237-9932

Contents

Do you hang your towel to dry in the sunshine after you go swimming? If so, you are using solar power! Heat or electricity that is made by capturing the Sun's **energy** is known as solar power. Solar power can heat swimming pools or keep greenhouses warm enough for plants to grow all year long. To change solar energy into electricity, we usually use **photovoltaic panels**. These panels can power **calculators**, homes, and spaceships.

Solar power is a renewable energy source. This means we cannot use it up. As long as the Sun shines, we can use its energy to power our world.

This field of photovoltaic panels is part of a solar power plant in Spain. The panels are angled so that they can collect the Sun's energy.

See the bright light. Feel the warm rays. Every day, Earth receives warmth and light from the Sun, our nearest star.

Have you ever wondered where the Sun gets its energy? The Sun is about one million times bigger than Earth. Because the Sun is so big, the **gravity** at its core, or center, is very strong. The Sun is made mostly of a gas called hydrogen. In the core, gravity and heat squash hydrogen together until it turns into **helium**. This **reaction** is called nuclear fusion. When the hydrogen changes into helium, it lets out a lot of energy. The energy beams out of the Sun as sunshine.

The distance between Earth and the Sun is 93 million miles (149.6 million km). However, it only takes 8.5 minutes for sunlight to reach Earth!

Have you ever noticed how hot cars get in the summer? This happens because the Sun's rays come in through car windows and get trapped inside. The trapped rays move around the car. This makes heat. Sometimes, people capture the Sun's energy on purpose. Greenhouses trap sunlight to make it warm enough for plants to grow. The energy we get by simply trapping the Sun's rays is called passive solar power.

Passive solar power can be used to heat water in pipes. It can also cook food in a solar oven. Some people even capture enough sunlight to heat their whole house.

Greenhouses usually have windows on both their roofs and walls to trap sunlight. The windows can be made from either plastic or glass.

People have always used the Sun for light and heat. However, it took some imagination to keep entire cities warm with solar power. About 2,500 years ago, builders in ancient Greece thought of a way to use the Sun's free energy. They built big cities with all the houses facing south, the sunniest direction. All winter, sunlight came in through the windows and warmed their houses.

In the American Southwest, native peoples, such as the Hopis, Pueblos, and Navajos, had the same idea. For 900 years, their homes have used passive solar power to provide sunlight and heat.

The Acoma Pueblos of New Mexico have lived in this passive solar city, called Sky City, for centuries. Even today, the city is not wired for electricity.

Passive solar power is not the only way we catch the Sun's energy. We also use photovoltaic panels to make electricity. These panels are made of thin pieces of **silicon**. Electricity moves easily through silicon.

Do you have a calculator? Most calculators have small photovoltaic panels across the top. Put your calculator in the sun. When light hits the panel, the calculator quickly powers up. You can use your calculator to add and subtract.

You can also often see photovoltaic panels on the roofs of houses and other buildings. The panels change sunlight into electricity that powers the buildings' lights, computers, TVs, and much more.

Some vineyards, or places where grapes are grown, use photovoltaic panels. The panels make electricity to power machines on the vineyard.

Solar Power Plants

Solar power plants can power whole cities! Some plants have many big photovoltaic panels that make electricity. Others, called **thermal** solar plants, use mirrors to aim sunlight at pipes of water or oil. Water becomes steam as it heats up, while oil is used to make steam from water. The steam moves to a **generator**, where it turns a **turbine**. This makes electricity.

Solar panels work well during the day. At night, though, there is no sunlight with which to make electricity. Solar plants must store energy. The plants sometimes use heat to warm up melted salt. The hot liquid salt is later used to make steam.

Parabolic trough plants, such as this one, are the most common kind of solar power plant. Their rows of curved mirrors aim sunlight at pipes.

Did you know that scientists have used **robots** to explore the surface of Mars? These robots were powered by solar panels. Solar panels also power **satellites** and space stations that **orbit** Earth.

Solar panels are often used to make electricity in hard to reach places. For example, scientists use solar panels to power up their tools in Antarctica, where there is no electricity. Some soldiers in the U.S. Army wear soft solar panels on their uniforms. They march and charge their gear at the same time. Solar panels allow us to take electricity almost anywhere!

This woman is clearing snow from solar panels on a mountain in Pakistan. The panels cannot make electricity if snow is covering them.

Solar power is a fairly clean way of making electricity. Making solar panels produces some pollution. However, once the panels are set up, they do not pollute. The sunshine on which solar panels run is free, and we will never run out of it.

However, solar power is not always useful because it does not work in the dark. People in the parts of the world close to the North Pole or the South Pole live in the dark for several months every year. There, solar power will work only part of the time. Solar panels can also be expensive to make and set up.

Families who put solar panels on their homes often save money on their energy bills. In some states, they can also save on their taxes.

Some scientists hope that soon we will have space-based solar power. Satellites in space could collect the Sun's energy. Then, **microwave** beams could send the energy down to a power plant. There, it could be changed into electricity for everyone to use. Since these satellites would be above any clouds, they would work on rainy days. We could use satellites to make electricity from solar energy any hour of the day and any day of the year.

Solar power offers many ways to make clean, renewable energy in the **future**. Maybe someday we will all drive solar-powered cars and talk on solar-powered cell phones!

Each year, students race solar cars in the American Solar Challenge. Students at the University of Missouri-Columbia built this solar car.

Solar Energy Timeline

4.5 billion years ago	The Sun forms from a cloud of hydrogen and helium.
400 BCE	The Greeks build passive solar cities.
1100 CE	The Pueblos build a passive solar city called Sky City in present-day New Mexico.
1839	The young French scientist Alexandre-Edmond Becquerel discovers how to make electricity from sunlight.
1891	Clarence Kemp becomes the first person to sell solar water heaters.
1954	**Engineers** at Bell Laboratories make photovoltaic panels. NASA, America's space agency, uses the panels to power satellites.
1979	U.S. president Jimmy Carter starts a national program to advance solar energy.
2007	The 70,000-panel Nellis Solar Power Plant opens in Nevada.
2008	Japanese scientists start building satellites to collect solar energy in space and beam it back to Earth.

Glossary

calculators (KAL-kyuh-lay-terz) Machines or tools used to do math.

energy (EH-nur-jee) The power to work or to act.

engineers (en-juh-NEERZ) Masters at planning and building engines, machines, roads, and bridges.

future (FYOO-chur) The time that is coming.

generator (JEH-neh-ray-tur) A machine that makes electricity.

gravity (GRA-vih-tee) The force of attraction between matter.

helium (HEE-lee-um) A light, colorless gas.

microwave (MY-kruh-wayv) Having to do with radio waves that have short wavelengths.

orbit (OR-bit) To travel a circular path.

photovoltaic panels (foh-toh-vol-TAY-ik PA-nelz) Flat objects that collect sunlight and change it into electricity.

reaction (ree-AK-shun) An action caused by something that happened.

robots (ROH-bots) Machines made to do jobs that often people do.

satellites (SA-tih-lyts) Spacecraft that circle Earth.

silicon (SIH-lih-kun) A kind of matter found in rocks and sand.

thermal (THER-mul) Using heat.

turbine (TER-byn) A motor that turns by a flow of water or air.

Index

Web Sites

Due to the changing nature of Internet links, PowerKids Press has developed an online list of Web sites related to the subject of this book. This site is updated regularly. Please use this link to access the list:
www.powerkidslinks.com/pow/solar/